GOD'S DESIGN® FOR HEAVEN & EARTH

OUR PLANET EARTH

TEACHER SUPPLEMENT

D1315439

1:1
answersingenesis
Petersburg, Kentucky, USA

ANSWERS IN GENESIS SCIENCE BY DEBBIE & RICHARD LAWRENCE

God's Design® for Heaven & Earth
Our Planet Earth Teacher Supplement

Second printing: July 2011

Published by Answers in Genesis, 2800 Bullittsburg Church Rd., Petersburg KY 41080

You may contact the authors at (970) 686-5744.

ISBN: 1-60092-294-5

Cover design & layout: Diane King
Editors: Lori Jaworski, Gary Vaterlaus

Printed in China.

www.answersingenesis.org   www.godsdesignscience.com

TABLE OF CONTENTS

WELCOME TO GOD'S DESIGN® FOR HEAVEN & EARTH

God's Design for Heaven and Earth is a series that has been designed for use in teaching earth science to elementary and middle school students. It is divided into three books: *Our Universe, Our Planet Earth,* and *Our Weather and Water.* Each book has 35 lessons including a final project that ties all of the lessons together.

In addition to the lessons, special features in each book include biographical information on interesting people as well as fun facts to make the subject more fun.

Although this is a complete curriculum, the information included here is just a beginning, so please feel free to add to each lesson as you see fit. A resource guide is included in the appendices to help you find additional information and resources. A list of supplies needed is included at the beginning of each lesson, while a master list of all supplies needed for the entire series can be found in the appendices.

Answer keys for all review questions, worksheets, quizzes, and the final exam are included here. Reproducible student worksheets and tests may be found on the supplementary CD-Rom for easy printing. Please contact Answers in Genesis if you wish to purchase a printed version of all the student materials, or go to www.AnswersBookstore.com.

If you wish to get through all three books of the *Heaven and Earth* series in one year, plan on covering approximately three lessons per week. The time required for each lesson varies depending on how much additional information you include, but plan on 20 minutes per lesson for beginners (grades 1–2) and 40 to 45 minutes for grades 3–8.

Quizzes may be given at the conclusion of each unit and the final exam may be given after lesson 34.

If you wish to cover the material in more depth, you may add additional information and take a longer period of time to cover all the material, or you could choose to do only one or two of the books in the series as a unit study.

WHY TEACH EARTH SCIENCE?

It is not uncommon to question the need to teach children hands-on science in elementary or middle school. We could argue that the knowledge gained in science will be needed later in life in order for children to be more productive and well-rounded adults. We could argue that teaching children science also teaches them logical and inductive thinking and reasoning skills, which are tools they will need to be more successful. We could argue that science is a necessity in this technological world in which we live. While all of these arguments are true, not one of them is the main reason that we should teach our children science. The most important reason to teach science in elementary school is to give children an understanding that God is our Creator, and the Bible can be trusted. Teaching science from a creation perspective is one of the best ways to reinforce our children's faith in God and to help them counter the evolutionary propaganda they face every day.

God is the Master Creator of everything. His handiwork is all around us. Our great Creator put in place all of the laws of physics, biology, and chemistry. These laws were put here for us to see His wisdom and power. In science, we see the hand of God at work more than in any other subject. Romans 1:20 says, "For since the creation of the world His invisible attributes are clearly seen, being understood by the things that are made, even His eternal power and Godhead, so that they [men] are without excuse." We need to help our children see God as Creator of the world around them so they will be able to recognize God and follow Him.

The study of earth science helps us to understand and appreciate this amazing world God gave us. Studying the processes that shape the earth, and exploring the origins of the earth and the universe often bring us into direct conflict with evolutionary theories. This is why it is so critical to teach our children the truth of the Bible, how to evaluate the evidence, how to distinguish fact from theory, and to realize that the evidence, rightly interpreted, supports biblical creation not evolution.

It's fun to teach earth science! It's interesting too. Rocks, weather, and stars are all around us. Children naturally collect rocks and gaze at the stars. You just need to direct their curiosity.

Finally, teaching earth science is easy. It's where you live. You won't have to try to find strange materials for experiments or do dangerous things to learn about the earth.

HOW DO I TEACH SCIENCE?

In order to teach any subject you need to understand how people learn. People learn in different ways. Most people, and children in particular, have a dominant or preferred learning style in which they absorb and retain information more easily.

If a student's dominant style is:

AUDITORY
He needs not only to hear the information but he needs to hear himself say it. This child needs oral presentation as well as oral drill and repetition.
VISUAL
She needs things she can see. This child responds well to flashcards, pictures, charts, models, etc.
KINESTHETIC
he needs active participation. This child remembers best through games, hands-on activities, experiments, and field trips.

Also, some people are more relational while others are more analytical. The relational student needs to know why this subject is important, and how it will affect him personally. The analytical student, however, wants just the facts.

If you are trying to teach more than one student, you will probably have to deal with more than one learning style. Therefore, you need to present your lessons in several different ways so that each student can grasp and retain the information.

GRADES 1–2

Because *God's Design Science* books are designed to be used with students in grades 1–8, each lesson has been divided into three sections. The "Beginner" section is for students in grades 1–2. This part contains a read-aloud section explaining the material for that lesson followed by a few questions to make sure that the students understand what they just heard. We recommend that you do the hands-on activity in the blue box in the main part of the lesson to help your students see and understand the concepts.

GRADES 3–8

The second part of each lesson should be completed by all upper elementary and junior high students. This is the main part of the lesson containing a reading section, a hands-on activity that reinforces the ideas in the reading section (blue box), and a review section that provides review questions and application questions (red box).

GRADES 6–8

Finally, for middle school/junior high age students, we provide a "Challenge" section that contains more challenging material as well as additional activities and projects for older students (green box).

We have included periodic biographies to help your students appreciate the great men and women who have gone before us in the field of science.

We suggest a threefold approach to each lesson:

INTRODUCE THE TOPIC

We give a brief description of the facts. Frequently you will want to add more information than the essentials given in this book. In addition to reading this section aloud (or having older children read it on their own), you may wish to do one or more of the following:

- Read a related book with your students.
- Write things down to help your visual learners.
- Give some history of the subject. We provide some historical sketches to help you, but you may want to add more.
- Ask questions to get your students thinking about the subject.

MAKE OBSERVATIONS AND DO EXPERIMENTS

- Hands-on projects are suggested for each lesson. This part of each lesson may require help from the teacher.
- Have your students perform the activity by themselves whenever possible.

REVIEW

- The "What did we learn?" section has review questions.
- The "Taking it further" section encourages students to
    - Draw conclusions
    - Make applications of what was learned
    - Add extended information to what was covered in the lesson
- The "FUN FACT" section adds fun or interesting information.

By teaching all three parts of the lesson, you will be presenting the material in a way that children with any learning style can both relate to and remember.

Also, this approach relates directly to the scientific method and will help your students think more scientifically. The *scientific method* is just a way to examine a subject logically and learn from it. Briefly, the steps of the scientific method are:

1. Learn about a topic.
2. Ask a question.
3. Make a hypothesis (a good guess).
4. Design an experiment to test your hypothesis.
5. Observe the experiment and collect data.
6. Draw conclusions. (Does the data support your hypothesis?)

Note: It's okay to have a "wrong hypothesis." That's how we learn. Be sure to help your students understand why they sometimes get a different result than expected.

Our lessons will help your students begin to approach problems in a logical, scientific way.

HOW DO I TEACH CREATION VS. EVOLUTION?

We are constantly bombarded by evolution-ary ideas about the earth in books, movies, museums, and even commercials. These raise many questions: What is the big bang? How old is the earth? Do fossils show evolution to be true? Was there really a worldwide flood? When did dinosaurs live? Was there an ice age? How can we teach our children the truth about the origins of the earth? The Bible answers these questions and this book accepts the historical accuracy of the Bible as written. We believe this is the only way we can teach our children to trust that everything God says is true.

There are five common views of the origins of life and the age of the earth:

Historical biblical account	Progressive creation	Gap theory	Theistic evolution	Naturalistic evolution
Each day of creation in Genesis is a normal day of about 24 hours in length, in which God created everything that exists. The earth is only thousands of years old, as determined by the genealogies in the Bible.	The idea that God created various creatures to replace other creatures that died out over millions of years. Each of the days in Genesis represents a long period of time (day-age view) and the earth is billions of years old.	The idea that there was a long, long time between what happened in Genesis 1:1 and what happened in Genesis 1:2. During this time, the "fossil record" was supposed to have formed, and millions of years of earth history supposedly passed.	The idea that God used the process of evolution over millions of years (involving struggle and death) to bring about what we see today.	The view that there is no God and evolution of all life forms happened by purely naturalistic processes over billions of years.

Any theory that tries to combine the evolutionary time frame with creation presupposes that death entered the world before Adam sinned, which contradicts what God has said in His Word. The view that the earth (and its "fossil record") is hundreds of millions of years old damages the gospel message. God's completed creation was "very good" at the end of the sixth day (Genesis 1:31). Death entered this perfect paradise *after* Adam disobeyed God's command. It was the punishment for Adam's sin (Genesis 2:16–17, 3:19; Romans 5:12–19). Thorns appeared when God cursed the ground because of Adam's sin (Genesis 3:18).

The first animal death occurred when God killed at least one animal, shedding its blood, to make clothes for Adam and Eve (Genesis 3:21). If the earth's "fossil record" (filled with death, disease, and thorns) formed over millions of years before Adam appeared (and before he sinned), then death no longer would be the penalty for sin. Death, the "last enemy" (1 Corinthians 15:26), diseases (such as cancer), and thorns would instead be part of the original creation that God labeled "very good." No, it is clear that the "fossil record" formed some time *after* Adam sinned—not many millions of years before. Most fossils were formed as a result of the worldwide Genesis Flood.

When viewed from a biblical perspective, the scientific evidence clearly supports a recent creation by God, and not naturalistic evolution and millions of years. The volume of evidence supporting the biblical creation account is substantial and cannot be adequately covered in this book. If you would like more information on this topic, please see the resource guide in Appendix A. To help get you started, just a few examples of evidence supporting biblical creation are given below:

Evolutionary Myth: The earth is 4.6 billion years old.

The Truth: Many processes observed today point to a young earth of only a few thousand years. The rate at which the earth's magnetic field is decaying suggests the earth must be less than 10,000 years old. The rate of population growth and the recent emergence of civilization suggests only a few thousand years of human population. And, at the current rate of accumulation, the amount of mud on the sea floor should be many kilometers thick if the earth were billions of years old. However, the average depth of all the mud in the whole ocean is less than 400 meters, giving a maximum age for the earth of not more than 12 million years. All this and more indicates an earth much younger than 4.6 billion years.

John D. Morris, *The Young Earth* (Creation Life Publishers, 1994), pp. 70–71, 83–90. See also "Get Answers: Young Age Evidence" at www.answersingenesis.org/go/young.

Evolutionary Myth: The universe formed from the big bang.

The Truth: There are many problems with this theory. It does not explain where the initial material came from. It cannot explain what caused that material to fly apart in the first place. And nothing in physics indicates what would make the particles begin to stick together instead of flying off into space forever. The big bang theory contradicts many scientific laws. Because of these problems, some scientists have abandoned the big bang and are attempting to develop new theories to explain the origin of the universe.

Jason Lisle, "Does the Big Bang Fit with the Bible," in *The New Answers Book 2*, Ken Ham, ed. (Master Books, 2008). See also "What are some of the problems with the big bang hypothesis?" at www.answersingenesis.org/go/big-bang.

Evolutionary Myth: Fossils prove evolution.

The Truth: While Darwin predicted that the fossil record would show numerous transitional fossils, even more than 145 years later, all we have are a handful of disputable examples. For example, there are no fossils showing something that is part way between a dinosaur and a bird. Fossils show that a snail has always been a snail; a squid has always been a squid. God created each animal to reproduce after its kind (Genesis 1:20–25).

Evolutionary Myth: There is not enough water for a worldwide flood.

The Truth: Prior to the Flood, just as today, much of the water was stored beneath the surface of the earth. In addition, Genesis 1 states that the water below was separated from the water above, indicating that the atmosphere may have contained a great deal more water than it does today. Also, it is likely that before the Flood the mountains were not as high as they are today, but that the mountains rose and the valleys sank *after* the Flood began, as Psalm 104:6–9 suggests. At the beginning of the Flood, the fountains of the deep burst forth and it rained for 40 days and nights. This could have provided more than enough water to flood the entire earth. Indeed, if the entire earth's surface were leveled by smoothing out the topography of not only the land surface but also the rock surface on the ocean floor, the waters of the present-day oceans would cover the earth's surface to a depth of 1.7 miles (2.7 kilometers). Fossils have been found on the highest mountain peaks around the world showing that the waters of the Flood did indeed cover the entire earth.

Ken Ham & Tim Lovett, "What There Really a Noah's Ark and Flood," in *The New Answers Book 1*, Ken Ham, ed. (Master Books, 2006).

Evolutionary Myth: Slow climate changes over time have resulted in multiple ice ages.

The Truth: There is widespread evidence of glaciers in many parts of the world indicating one ice age. Evolutionists find the cause of the Ice Age a mystery. Obviously, the climate would need to be colder. But global cooling by itself is not enough, because then there would be less evaporation, so less snow. How is it possible to have both a cold climate and lots of evaporation? The Ice Age was most likely an aftermath of Noah's Flood. When "all the fountains of the great deep" broke up, much hot water and lava would have poured directly into the oceans. This would have warmed the oceans, increasing evaporation. At the same time, much volcanic ash in the air after the Flood would have blocked out much sunlight, cooling the land. So the Flood would have produced the necessary combination of increased evaporation from the warmed oceans and cool continental climate from the volcanic ash in the air. This would have resulted in increased snowfall over the continents. With the snow falling faster than it melted, ice sheets would have built up. The Ice Age probably lasted less than 700 years.

Michael Oard, *Frozen in Time* (Master Books, 2004). See also www.answersingenesis.org/go/ice-age.

Evolutionary Myth: Thousands of random changes over millions of years resulted in the earth we see today.

The Truth: The second law of thermodynamics describes how any system tends toward a state of zero entropy or disorder. We observe how everything around us becomes less organized and loses energy. The changes required for the formation of the universe, the planet earth and life, all from disorder, run counter to the physical laws we see at work today. There is no known mechanism to harness the raw energy of the universe and generate the specified complexity we see all around us.

John D. Morris, *The Young Earth* (Creation Life Publishers, 1994), p. 43. See also www.answersingenesis.org/go/thermodynamics.

Despite the claims of many scientists, if you examine the evidence objectively, it is obvious that evolution and millions of years have not been proven. You can be confident that if you teach that what the Bible says is true, you won't go wrong. Instill in your student a confidence in the truth of the Bible in all areas. If scientific thought seems to contradict the Bible, realize that scientists often make mistakes, but God does not lie. At one time scientists believed that the earth was the center of the universe, that living things could spring from non-living things, and that blood-letting was good for the body. All of these were believed to be scientific facts but have since been disproved, but the Word of God remains true. If we use modern "science" to interpret the Bible, what will happen to our faith in God's Word when scientists change their theories yet again?

Integrating the Seven C's

Throughout the *God's Design® for Science* series you will see icons that represent the Seven C's of History. The Seven C's is a framework in which all of history, and the future to come, can be placed. As we go through our daily routines we may not understand how the details of life connect with the truth that we find in the Bible. This is also the case for students. When discussing the importance of the Bible you may find yourself telling students that the Bible is relevant in everyday activities. But how do we help the younger generation see that? The Seven C's are intended to help.

The Seven C's can be used to develop a biblical worldview in students, young or old. Much more than entertaining stories and religious teachings, the Bible has real connections to our everyday life. It may be hard, at first, to see how many connections there are, but with practice ,the daily relevance of God's Word will come alive. Let's look at the Seven C's of History and how each can be connected to what the students are learning.

Creation

God perfectly created the heavens, the earth, and all that is in them in six normal-length days around 6,000 years ago.

This teaching is foundational to a biblical worldview and can be put into the context of any subject. In science, the amazing design that we see in nature—whether in the veins of a leaf or the complexity of your hand—is all the handiwork of God. Virtually all of the lessons in *God's Design for Science* can be related to God's creation of the heavens and earth.

Other contexts include:

Natural laws—any discussion of a law of nature naturally leads to God's creative power.

DNA and information—the information in every living thing was created by God's supreme intelligence.

Mathematics—the laws of mathematics reflect the order of the Creator.

Biological diversity—the distinct kinds of animals that we see were created during the Creation Week, not as products of evolution.

Art—the creativity of man is demonstrated through various art forms.

History—all time scales can be compared to the biblical time scale extending back about 6,000 years.

Ecology—God has called mankind to act as stewards over His creation.

Corruption

After God completed His perfect creation, Adam disobeyed God by eating the forbidden fruit. As a result, sin and death entered the world, and the world has been in decay since that time. This point is evident throughout the world that we live in. The struggle for survival in animals, the death of loved ones, and the violence all around us are all examples of the corrupting influence of sin.

Other contexts include:

Genetics—the mutations that lead to diseases, cancer, and variation within populations are the result of corruption.

Biological relationships—predators and parasites result from corruption.

History—wars and struggles between mankind, exemplified in the account of Cain and Abel, are a result of sin.

CATASTROPHE

God was grieved by the wickedness of mankind and judged this wickedness with a global Flood. The Flood covered the entire surface of the earth and killed all air-breathing creatures that were not aboard the Ark. The eight people and the animals aboard the Ark replenished the earth after God delivered them from the catastrophe.

The catastrophe described in the Bible would naturally leave behind much evidence. The studies of geology and of the biological diversity of animals on the planet are two of the most obvious applications of this event. Much of scientific understanding is based on how a scientist views the events of the Genesis Flood.

Other contexts include:

Biological diversity—all of the birds, mammals, and other air-breathing animals have populated the earth from the original kinds which left the Ark.

Geology—the layers of sedimentary rock seen in roadcuts, canyons, and other geologic features are testaments to the global Flood.

Geography—features like mountains, valleys, and plains were formed as the floodwaters receded.

Physics—rainbows are a perennial sign of God's faithfulness and His pledge to never flood the entire earth again.

Fossils—Most fossils are a result of the Flood rapidly burying plants and animals.

Plate tectonics—the rapid movement of the earth's plates likely accompanied the Flood.

Global warming/Ice Age—both of these items are likely a result of the activity of the Flood.

The warming we are experiencing today has been present since the peak of the Ice Age (with variations over time).

CONFUSION

God commanded Noah and his descendants to spread across the earth. The refusal to obey this command and the building of the tower at Babel caused God to judge this sin. The common language of the people was confused and they spread across the globe as groups with a common language. All people are truly of "one blood" as descendants of Noah and, originally, Adam.

The confusion of the languages led people to scatter across the globe. As people settled in new areas, the traits they carried with them became concentrated in those populations. Traits like dark skin were beneficial in the tropics while other traits benefited populations in northern climates, and distinct people groups, not races, developed.

Other contexts include:

Genetics—the study of human DNA has shown that there is little difference in the genetic makeup of the so-called "races."

Languages—there are about seventy language groups from which all modern languages have developed.

Archaeology—the presence of common building structures, like pyramids, around the world confirms the biblical account.

Literature—recorded and oral records tell of similar events relating to the Flood and the dispersion at Babel.

CHRIST

God did not leave mankind without a way to be redeemed from its sinful state. The Law was given to Moses to show how far away man is from God's standard of perfection. Rather than the sacrifices, which only covered sins, people needed a Savior to take away their sin. This was accomplished when Jesus Christ came to earth to live a perfect life and, by that obedience, was able to be the sacrifice to satisfy God's wrath for all who believe.

The deity of Christ and the amazing plan that was set forth before the foundation of the earth is the core of Christian doctrine. The earthly life of Jesus was the fulfillment of many prophecies and confirms the truthfulness of the Bible. His miracles and presence in human form demonstrate that God is both intimately concerned with His creation and able to control it in an absolute way.

Other contexts include:

Psychology—popular secular psychology teaches of the inherent goodness of man, but Christ has lived the only perfect life. Mankind needs a Savior to redeem it from its unrighteousness.

Biology—Christ's virgin birth demonstrates God's sovereignty over nature.

Physics—turning the water into wine and the feeding of the five thousand demonstrate Christ's deity and His sovereignty over nature.

History—time is marked (in the western world) based on the birth of Christ despite current efforts to change the meaning.

Art—much art is based on the life of Christ and many of the masters are known for these depictions, whether on canvas or in music.

CROSS

Because God is perfectly just and holy, He must punish sin. The sinless life of Jesus Christ was offered as a substitutionary sacrifice for all of those who will repent and put their faith in the Savior. After His death on the Cross, He defeated death by rising on the third day and is now seated at the right hand of God.

The events surrounding the crucifixion and resurrection have a most significant place in the life of Christians. Though there is no way to scientifically prove the resurrection, there is likewise no way to prove the stories of evolutionary history. These are matters of faith founded in the truth of God's Word and His character. The eyewitness testimony of over 500 people and the written Word of God provide the basis for our belief.

Other contexts include:

Biology—the biological details of the crucifixion can be studied alongside the anatomy of the human body.

History—the use of crucifixion as a method of punishment was short-lived in historical terms and not known at the time it was prophesied.

Art—the crucifixion and resurrection have inspired many wonderful works of art.

CONSUMMATION

God, in His great mercy, has promised that He will restore the earth to its original state—a world without death, suffering, war, and disease. The corruption introduced by Adam's sin will be removed. Those who have repented and put their trust in the completed work of Christ on the Cross will experience life in this new heaven and earth. We will be able to enjoy and worship God forever in a perfect place.

This future event is a little more difficult to connect with academic subjects. However, the hope of a life in God's presence and in the absence of sin can be inserted in discussions of human conflict, disease, suffering, and sin in general.

Other contexts include:

History—in discussions of war or human conflict the coming age offers hope.

Biology—the violent struggle for life seen in the predator-prey relationships will no longer taint the earth.

Medicine—while we struggle to find cures for diseases and alleviate the suffering of those enduring the effects of the Curse, we ultimately place our hope in the healing that will come in the eternal state.

The preceding examples are given to provide ideas for integrating the Seven C's of History into a broad range of curriculum activities. We would recommend that you give your students, and yourself, a better understanding of the Seven C's framework by using AiG's *Answers for Kids* curriculum.

The first seven lessons of this curriculum cover the Seven C's and will establish a solid understanding of the true history, and future, of the universe. Full lesson plans, activities, and student resources are provided in the curriculum set.

We also offer bookmarks displaying the Seven C's and a wall chart. These can be used as visual cues for the students to help them recall the information and integrate new learning into its proper place in a biblical worldview.

Even if you use other curricula, you can still incorporate the Seven C's teaching into those. Using this approach will help students make firm connections between biblical events and every aspect of the world around them, and they will begin to develop a truly biblical worldview and not just add pieces of the Bible to what they learn in "the real world."

Unit 1
ORIGINS & GLACIERS

LESSON 1

INTRODUCTION TO EARTH SCIENCE

THE STUDY OF OUR WORLD

SUPPLY LIST

Tennis ball   String   Masking tape

BEGINNERS

- What is earth science? **The study of the earth.**
- Where did the earth come from? **God created it.**
- What other things did God create? **The sun, moon, stars, sky, land, plants, animals, people.**

WHAT DID WE LEARN?

- What are the four main studies of earth science? **Space/astronomy, atmosphere/meteorology, lithosphere/geology, and water/hydrology.**
- What is one question mentioned in this lesson that science cannot answer about the earth? **Where it came from originally. There are many other questions beyond science as well.**
- Why can we rely on God's Word to tell us where the earth came from? **The Bible is the Word of God and God does not lie. The evidence around us confirms what the Bible says.**

TAKING IT FURTHER

- How does the first law of thermodynamics confirm the Genesis account of creation? **Since energy cannot be created by natural means, then a supernatural event must have occurred to make the energy and matter in the universe. The Bible says God spoke it into existence.**
- How does the second law of thermodynamics confirm the Genesis account of creation? **If the whole universe is slowing down and losing energy, there must have been a time when everything was started—when the energy was put into the whole system. Together with the first law, this shows that a supernatural event, such as the creation described in the Bible, must have happened in the past.**
- Read Psalm 139:8–10. What do these verses say about where we can find God? **God is present everywhere, even in space or at the bottom of the ocean; wherever we go, God is with us.**

LESSON 2

INTRODUCTION TO GEOLOGY

THE STUDY OF THE EARTH ITSELF

SUPPLY LIST

Copy of "Geology Scavenger Hunt"
Supplies for Challenge: "Periodic Table of the Elements"(page 15 in student manual)
Packaged food with nutrition labels

BEGINNERS

- What is geology? **The study of the earth.**

- What makes earth a special planet? **It has lots of water, is the right distance from the sun.**

- What are some things around you that came from the earth? **Anything made from metal, plastic, or rock. Anything made from plants because they grow in the soil.**

GEOLOGY SCAVENGER HUNT

1. Milk	2. Toothpaste	3. Salt	4. Cereal
5. Pencils	6. Chalk	7. Matches	8. Baby powder
9. Rechargable batteries	10. Drywall	11. Computer chips	12. Pennies
13. Electrical wiring	14. Thermometer	15. Paper clip or staple	

WHAT DID WE LEARN?

- What is geology? **The study of the earth and the processes that affect it.**

- What are some of the evidences that God designed the earth uniquely to support life? **The abundance of water and its properties, just the right amount of oxygen in the atmosphere, the distance of the earth from the sun, the tilt of the earth.**

TAKING IT FURTHER

- List some ways that geology affects your life on a regular basis. **Minerals and metals are in nearly every object around you. You use gas in your car. Your house must be built on a firm foundation.**

- What area of geology interests you the most? **Go to the library and learn more about it.**

CHALLENGE: ELEMENTS

- **You are likely to find sodium, potassium, phosphorus, magnesium, zinc, copper, iron, and calcium.**

LESSON 3

THE EARTH'S HISTORY

HOW IT ALL BEGAN

SUPPLY LIST

Jar with lid   Rocks, pebbles, sand, dirt   Water

BEGINNERS

- What are the three important events that made the earth the way it is today? **Creation, Fall, Flood.**
- Does the Bible indicate that the earth is thousands or billions of years old? **The Bible teaches thousands of years.**

WHAT DID WE LEARN?

- What are the two most popular views for how the earth became what it is today? **Creation/biblical and evolution/uniformitarian.**
- According to the Bible, what are the three major events that affected the way the earth looks today? **Creation, the Fall of man, the Flood.**   why?
- Should a good scientist disregard evidence that contradicts his/her ideas? **No, he/she should examine the evidence and try to understand why it does not agree. Sometimes the answer cannot be found now, but may be obvious later, when other discoveries are made.**
- Have scientists proven that evolution is true? **No! Evolution is a model of origins that cannot be proven, and actually contradicts what we observe. A historical event, such as the origin of the world, cannot be recreated or tested, so it cannot be proven. We must trust the account of the One who was there—God.**
- Have scientists proven that biblical creation is true? **No. Creation and evolution both deal with historical events—origins science. Neither can be proven by science. But, when the evidence is examined, it contradicts the evolutionary view and confirms the Bible's account.**

TAKING IT FURTHER

- How might scientists explain the discovery of fossilized seashells in the middle of a desert? **An ocean must have covered the desert at one time. The Bible says the whole world was covered with water during the Flood. Evolutionists say the climate was very different in the past, causing the oceans to cover more of the world.**
- Explain how a fossilized tree could be found upright through several layers of rock. **The tree had to have been covered with the various layers before it had a chance to decay. This must have occurred relatively quickly, over a few years time, not over millions of years. In fact, we see an example where hundreds of trees are settling upright into sediment at the bottom of Spirit Lake following the 1980 eruption of Mount St. Helens.**

THE GREAT FLOOD

GOD'S PUNISHMENT FOR SIN

SUPPLY LIST

Paper   Drawing materials (colored pencils, markers, etc.)
Supplies for Challenge: Copy of "Did the Flood Really Happen?" worksheet

BEGINNERS

- Why did God send the Genesis Flood? **To punish man's wickedness.**
- How did the Flood affect the surface of the earth? **It laid down layers of mud that formed new rocks, and fossils, and it washed away areas of the earth.**
- How was the weather different after the Flood? **It was probably much cooler.**

WHAT DID WE LEARN?

- What are some things geologists observe that point to a worldwide Flood? **Large amounts of sedimentary rock, abundant fossils, most fossils are aquatic.**
- What major geological events may have been associated with the Flood of Noah's day? **Major volcanic activity, separation of the landmasses, the Ice Age, formation of mountain ranges.**

TAKING IT FURTHER

- How would a huge flood change the way the earth looks? **Rushing water would cause massive erosion, wearing away rock. This would cause valleys to form and would move rock and soil from one place to another. It would also bury massive amounts of plants and animals under thick layers of mud resulting in abundant fossils, as well as coal and oil deposits.**
- Why did God send a huge flood? **To punish man for his wickedness.**

CHALLENGE: DID THE FLOOD REALLY HAPPEN? WORKSHEET

1. _**Yes**_ Water would wash away rocks, soil, and buildings.
2. _**Yes**_ Millions of animals and people would die.
3. _**Yes**_ The water would move rocks and soil from one place to another.
4. _**Yes**_ Buildings would be destroyed.
5. _**No**_ Land would be unchanged.
6. _**Yes**_ A large boat would float on the water.
7. _**Yes**_ Land animals would be covered with mud and sand.
8. _**Yes**_ Debris would settle out of the water.
9. _**No**_ Plants would not be uprooted or killed.
10. _**Yes**_ New paths would be formed for water to flow through.

With this in mind, list at least three things that you would expect to find, hundreds or even thousands of years later when you dig into the earth.

- **Millions of fossils—mostly of sea creatures found all over the world.**
- **Fossils of sea creatures on the tops of high mountains.**

- **Fossils of sea creatures in the middle of deserts.**
- **Oil and coal that were formed from dead plants and sea creatures that were buried under tons of rock.**
- **Layers of sedimentary rock.**
- **Some rock layers are curved or folded as if they were all soft at the same time.**
- **Deep canyons were carved.**

These are just a few evidences that support the idea of a worldwide flood.

LESSON 5

THE GREAT ICE AGE

THE AGE OF WOOLY MAMMOTHS

SUPPLY LIST

Copy of "Ice Age Crossword Puzzle"

Supplies for Challenge: Copy of world map   World Atlas with climate map of world

BEGINNERS

- How did the weather after the Flood compare to the weather before the Flood? **The weather was probably much cooler.**
- What could make the weather cooler after the Flood? **More clouds and more ash from volcanoes.**
- Was the earth completely covered with ice during the Ice Age? **No, much of the earth was still warm and people lived there.**

ICE AGE CROSSWORD PUZZLE

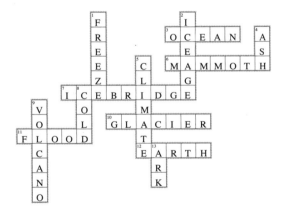

WHAT DID WE LEARN?

- What two conditions are necessary for an ice age? **Wetter winters and cooler summers.**
- How did the Genesis Flood set up conditions for a great ice age? **Warm oceans allowed for lots of evaporation and therefore lots of snow. Volcanic ash blocked much of the sunlight, causing much cooler summers.**
- How do evolutionists explain the needed conditions for multiple ice ages? **They cannot adequately explain what causes the additional moisture or the cooler temperatures.**

- Do we see new glaciers forming today? **We see some new glaciers forming and some old glaciers growing bigger for a few years, but not on the large scale they did during the Ice Age.**

LESSON 6

GLACIERS

ICE THAT NEVER MELTS

SUPPLY LIST

Several ice cubes   Dish of water   Toy boat that will fit in the dish

BEGINNERS

- What is a glacier? **A sheet of ice that does not completely melt even in the summer.**
- Where are most glaciers located? **At the North and South Poles.**
- Where do icebergs come from? **They are large pieces of ice that have broken off of glaciers and fallen into the water.**

WHAT DID WE LEARN?

- What is a glacier? **A formation of ice that does not completely melt from year to year.**
- How does a glacier form? **The snow accumulates each year because it does not completely melt during the summer. The weight of the snow compacts the snow below it, eventually turning it to ice.**
- What are the three types of glaciers? **Valley—forms in a valley, Piedmont—spreads out from two or more valleys, Continental—forms in a relatively flat area and spreads out in all directions.**
- What is calving? **When glaciers reach water and pieces break off into the water.**

TAKING IT FURTHER

- Why do glaciers exist mostly at the poles and on high mountain tops? **That is where it stays cold enough in the summers to keep the snow and ice from completely melting.**
- Why is it cold enough to prevent glaciers from melting at the North Pole, when there is 20–24 hours of sunlight during the summer? **Even though there are many hours of sunlight, the light hits the earth at a steep angle, so most of the heat is reflected.**

LESSON 7

MOVEMENT OF GLACIERS

SLOWLY CREEPING DOWN THE VALLEY

SUPPLY LIST

Empty half-gallon milk carton   Water, sand, and pebbles   Pair of gloves

Note: this activity requires the water to freeze overnight. You may want to start it the day before you do the lesson.

Supplies for Challenge: Glass jar with lid   Newspaper   Plastic zipper bag   Work gloves

BEGINNERS

- What makes a glacier move? **Gravity.**

- What moves along with the ice? **Dirt, rocks, other debris.**

- How can you tell how far a glacier has moved in the past? **Rocks and dirt that were pushed by the glacier are left in a line ahead of the glacier's farthest position.**

WHAT DID WE LEARN?

- What is the shape of a valley carved by glaciers? **U-shaped.**

- How do glaciers pick up rocks and other debris? **They melt, and the water flows around the rocks; then the water refreezes and the rocks become part of the glacier.**

- What is the name of the line of rocks that marks the furthest advance of the glacier? **Terminal moraine.**

TAKING IT FURTHER

- How might a scientist tell how far a glacier moved a rock or boulder? **One way is to test what kind of rock the boulder is made of. Often, it is a different type than the rocks around it. Then, the scientist must trace the path of the glacier backward to where that type of rock is found.**

- Why do glaciers often have deep cracks and crevices? **The lower layers of ice move smoothly while the upper layers, which are less compressed, are more brittle and break rather than move with the glacier.**

QUIZ
1

ORIGINS & GLACIERS

LESSONS 1–7

Mark each statement as either True or False.

1. _T_ We can rely on the Bible to tell us the truth about God and His creation.

2. _F_ We can prove scientifically where the earth came from.

3. _F_ Science can answer all of our questions.

4. _T_ Fossils have been located in every part of the world.

5. _T_ The biblical account of the Flood explains much of what we see on earth.

6. _F_ A scientist should disregard evidence that contradicts his/her theories.

7. _T_ Scientists have not proven evolution to be true.

8. _T_ The abundance of aquatic fossils is consistent with a worldwide Flood.

9. _T_ The worldwide Flood was God's punishment for man's sin.

10. _T_ Evolutionists cannot adequately explain how conditions formed to create an ice age.

Short answer:

11. List three biblical events that greatly affected the surface of the earth. **Creation, Fall, Flood.**

12. Describe three attributes of the earth that make it just right for life to occur here.

Distance from sun, tilt of axis, properties and abundance of water, oxygen/nitrogen ratio, size provides just the right amount of gravity.

13. List three ways that geology affects your life. **Minerals in food, anything made of metal, oil, anything made of plastic, soil to grow plants, caves to explore, etc.**

14. List the two climate conditions required for an ice age. **Wet winters and cool summers.**

15. List the four main studies of earth science. **Astronomy, meteorology, geology, hydrology.**

CHALLENGE QUESTIONS

Short answer:

16. Explain what the following quote is saying about scientists who believe in evolution.

    Dr. Scott Todd, an immunologist at Kansas State University: "Even if all the data point to an intelligent designer, such an hypothesis is excluded from science because it is not naturalistic." **Evolutionists will ignore the data if they point to a Creator.**

17. Based on what you have learned about the great Ice Age, in which areas would you expect to see evidence of glaciers? Write yes if you would expect to see it and no if you would not expect to see it.

    A. _Yes_ Canada   B. _Yes_ Montana   C. _No_ Mexico

    D. _Yes_ Norway   E. _Yes_ Siberia   F. _No_ Egypt

18. List three economic or social effects caused by the Little Ice Age. **Changes in fishing grounds; changes in the crops that could be grown; some people had to move as glaciers advanced; less food was available so some people suffered from famine, etc.**

ROCKS & MINERALS

LESSON 8

DESIGN OF THE EARTH

BLUEPRINT FOR THE PLANET

SUPPLY LIST

1 gumball per child   Bowl   1 large marshmallow per child   Pencil
Chocolate chips   Small bottle or jar   1 toothpick per child   Waxed paper
Supplies for Challenge: Chocolate chips   Plastic zipper bag   Cup   Water

BEGINNERS

- What are the three parts of the earth? **The crust, mantle, and core.**
- What is the hottest part of the earth? **The core.**
- What is the thinnest part of the earth? **The crust.**

WHAT DID WE LEARN?

- What do most scientists believe to be the three main parts of the earth? **The core, the mantle and the crust.**
- Which is the thickest part of the earth? **The mantle.**
- Which is the thinnest part of the earth? **The crust.**
- Where is the crust the thickest? **Under the mountains.**

TAKING IT FURTHER

- Why do scientists believe the mantle is hotter and denser than the crust? **Earthquake or seismic waves travel more quickly through the mantle than through the crust.**
- For what other things, besides the interior of the earth, do scientists have to develop models without actually seeing what they are describing? **Very small things like atoms, very large things like the universe, things they are designing like airplanes or space ships.**

LESSON 9

ROCKS

BOULDERS, ROCKS, GRAVEL, PEBBLES . . .

SUPPLY LIST

Copy of "The Rock Cycle" worksheet
Supplies for Challenge: Several rocks

BEGINNERS

- Where can you find rocks? **Everywhere.**
- What are the three different kinds of rocks? **Igneous, sedimentary, metamorphic.**

THE ROCK CYCLE WORKSHEET

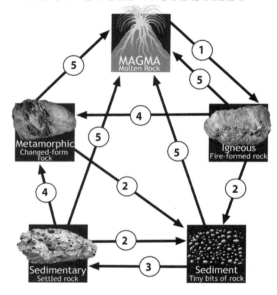

WHAT DID WE LEARN?

- What are rocks made from? **One or more minerals or organic materials.**
- What are the three categories of rocks? **Igneous, sedimentary, and metamorphic.**
- How is igneous rock formed? **Igneous rock forms when melted rock, called magma, cools.**
- How is sedimentary rock formed? **Sedimentary rock forms when layers of sediment are pressed and cemented together in some way.**
- How is metamorphic rock formed? **Metamorphic rock is formed when either igneous or sedimentary rock is exposed to high pressure and high temperature for an extended period of time. The rock's atomic structure is changed and it becomes a different type of rock.**

TAKING IT FURTHER

- Why are rocks important? **They affect nearly every area of our lives.**

- Where is a good place to look for rocks? **You can find rocks nearly everywhere. It depends on what specific kinds of rocks you are looking for. You will more easily find rocks in areas with little or no soil.**
- Why is it better to store your rock samples in a box with dividers than in a bag? **Some rocks are harder than others. In a bag, the samples will hit against each other and may scratch or break each other.**

LESSON 10

IGNEOUS ROCKS

FIRE ROCKS

SUPPLY LIST

Alum   Saucepan   Water   2 craft sticks   2 paper or plastic cups
Supplies for Challenge: Several samples of igneous rocks   Rock and Mineral Guide
Magnifying glass

BEGINNERS

- How is igneous rock formed? **When liquid rock cools it forms igneous rocks.**
- Do crystals grow in rocks that formed inside the earth or on top of the earth? **Inside.**
- What are two uses for igneous rocks? **Monuments and arrow heads.**

WHAT DID WE LEARN?

- What is the difference between magma and lava? **Magma is inside the earth's crust and lava is on top of the earth's crust.**
- How are extrusive rocks formed? **When magma flows to an area where it cools quickly, extrusive rocks are formed. This happens when lava flows to the surface of the earth's crust.**
- How are intrusive rocks formed? **When magma flows to an area inside the crust that is cool enough for it to solidify, but at a slow rate, it forms intrusive rocks.**

TAKING IT FURTHER

- Which kind of igneous rocks have the largest crystals? **The ones that were formed slowly—intrusive rocks.**
- Why is granite commonly used in buildings and monuments? **It is hard and strong, and can be polished.**
- Do all rocks sink in water? **No, pumice often floats.**
- Why not? **Some rocks have air holes in them making them less dense than water, so they float.**
- Where are you likely to find pumice? **Near a previously active volcano.**

LESSON 11

SEDIMENTARY ROCKS

LAYERS OF SEDIMENT

SUPPLY LIST

2 cups of sand   1 cup of cornstarch   Old saucepan   Paint (optional)
Optional: Ingredients to make a peanut butter and jelly or lunchmeat sandwich
Supplies for Challenge: 2 Paper cups   Plaster of Paris   Smooth pebbles
Rough pebbles (like aquarium rocks)   Rock and Mineral Guide

BEGINNERS

- How is a sedimentary rock made in nature? **Water moves bits of rock, sand, or other things around and when they settle they are glued together by chemicals in the water. Then they dry into rocks as the water dries up.**
- When were many of the sedimentary rocks formed? **During the Genesis Flood.**
- Name one common sedimentary rock. **Sandstone, limestone, or man-made rocks such as bricks or sidewalk chalk.**

WHAT DID WE LEARN?

- How are sedimentary rocks formed? **When fragments of rocks and other debris settle out of water they form strata or layers. These layers are pressed together and glued with natural cement to form fragmental sedimentary rock. Chemical rock forms as chemicals precipitate from the water and harden into rock.**
- Were all sedimentary rocks formed during the Flood? **No, but much of the sedimentary rock was probably formed then. New sedimentary rock is still being formed today, especially in caves and other wet areas.**

TAKING IT FURTHER

- Why are fossils found in sedimentary rocks? **Fossils form when plants or animals are covered over with mud or other sediment shortly after dying. Sea creatures were covered over with layers of sediment during the Flood and these layers formed into much of the sedimentary rock we find today. Thus, most of the fossils are found in these rocks.**
- Sediment is simply any small piece of something that settles out of a liquid. What sediment might you find around your house or in nature?
  Coffee grounds, tea leaves, and orange juice pulp are a few sediments you might find around your house. Silt is a common sediment found in lakes and ponds. Glaciers melt and leave behind pebbles, rocks and even boulders as sediment.

FOSSILS

HOW DO WE KNOW WHAT DINOSAURS LOOKED LIKE?

SUPPLY LIST

Plaster of Paris   Modeling clay   Petroleum jelly   Cup and spoon
Seashell or other item to "fossilize"

BEGINNERS

- What kinds of animals are most fossils? **Sea creatures such as clams and snails.**

- Why don't most animals become fossils? **Because they are not covered quickly with mud or sand when they die so they quickly decay.**

- What kind of rock contains fossils? **Sedimentary.**

WHAT DID WE LEARN?

- How does an animal become a fossil? **If it is covered by mud, wet sand or other similar substance shortly after is dies, instead of decaying, the hard structures can slowly be replaced by minerals.**

- What are the two different types of fossils? **Cast fossils—imprints; Mold fossils—rock structures in the shape of the original structure.**

- What types of creatures are most fossils? **95% of all fossils are sea invertebrates such as clams and other shellfish.**

TAKING IT FURTHER

- How many true transitional fossils, ones showing one creature evolving into another, have been found? **None!**

- What does this indicate about the theory that land animals evolved from sea creatures? **It shows that there is no physical evidence to support that theory.**

- What are some things we can learn from fossils? **What kinds of plants and animals existed in the past, how those plants and animals were shaped compared to species that exist today, and what plants and animals are now extinct.**

- What kinds of things cannot be learned from fossils? **Usually soft structures such as noses, hair, etc. are not preserved, so it is hard to tell exactly what a creature looked like. You can't tell its color either. Also, if only one or two bones are found, it is very difficult to know with certainty what that creature looked like. Some characteristics can be implied from fossil evidence but cannot be proven. For example, many scientists believe that at least some of the dinosaurs were warm-blooded because the distance between fossilized footprints indicates they must have been able to move quickly—an ability usually limited to warm-blooded creatures. However, this characteristic cannot be proven from the fossil evidence alone.**

LESSON 13

FOSSIL FUELS

A MAJOR ENERGY SOURCE

SUPPLY LIST

2 sponges   Epsom salt   Shallow dish or pan   Scissors   Food coloring (optional)

Supplies for Challenge: Copy of "What Would You Expect?" worksheet

BEGINNERS

- What are the three types of fossil fuels? **Coal, oil, natural gas.**

- What is coal made from? **Plants.**

- What is oil made from? **Sea creatures.**

- What do we use fossil fuels for? **To generate electricity, to make gasoline for our cars, to heat our homes, etc.**

WHAT DID WE LEARN?

- What is the definition of a fossil fuel? **Fuel that was formed from dead plants or animals that were changed by heat and pressure over time.**

- What three forms of fossil fuels do we commonly use? **Coal, oil/petroleum, and natural gas.**

TAKING IT FURTHER

- What evidence supports rapid and recent coal formation instead of slow formation millions of years ago? **Carbon-14 dating, in spite of its limitations, supports recent formation. Boulders and delicate fossils in coal beds also show a young age.**

- Why is finding natural gas when drilling into the ground a good indicator that oil is nearby? **Natural gas is believed to be a by-product of oil formation and is often found in the rocks surrounding an oil field.**

- Why is the existence of natural gas an indication that oil was formed only a few thousand years ago? **Natural gas is a by-product of oil production. If the oil was formed millions of years ago, the gas would have escaped through the rocks by now.**

CHALLENGE: WHAT WOULD YOU EXPECT? WORKSHEET

Lowest level	Upper level	Not found
Flies	**Pine needles**	**Palm trees**
Fish	**Flies**	**Coconuts**
Algae	**Deer**	**Sea shells**
Crawdads	**Scrub oak**	**Parrots**
Possibly wild flowers and pine needles	**Wild flowers**	**Possibly crows**
	Ground squirrel	
	Snakes	
	Rabbits	
	Possibly crows	

Accept reasonable answers.

- Briefly explain why you put the items where you did. **You would not expect to find items that do not naturally live in the Rocky Mountains such as palm trees, coconuts, sea creatures and parrots. Also, birds could possibly fly to safety and may not have been trapped by the eruption. You would expect to find plants and animals that live in and near the lake in a lower level than those that live above the lake, although some items, such as flowers or pine needles may have been pushed or blown down the hill into the lake. Overall you would expect the land plants and animals to be found above the water plants and animals. Similarly, if there was a worldwide Flood that lasted for five months, you would expect the creatures that lived near the bottom of the ocean to be found in lower layers than those that lived in higher land areas.**

LESSON 14

METAMORPHIC ROCKS

LET'S MAKE A CHANGE

SUPPLY LIST

Copy of "Morphing Ice" worksheet   Waxed paper   Sauce pan
Shaved ice (or snow if available)   2 or 3 pieces of different colored taffy or other soft candy
Supplies for Challenge: Copy of "Metamorphic Match" worksheet   Rock and Mineral Guide

BEGINNERS

- What does the word metamorphic mean? **Changed.**
- What does a rock have to experience to become a changed rock? **Heat, pressure, and time.**
- Name one kind of metamorphic rock? **Marble.**

MORPHING ICE WORKSHEET

Heat only: the rock melts and becomes magma; Pressure only: the rock breaks into smaller pieces; Time only: nothing happens to the rock—it remains the same. Heat, pressure, and time: the rock is changed into metamorphic rock as the molecular structure of the rock changes.

WHAT DID WE LEARN?

- What are the three ingredients needed to change igneous or sedimentary rock into metamorphic rock? **Heat, pressure, and time.**
- Why is marble often swirled instead of pure white? **There are often impurities in the limestone deposits that are a different color from the limestone.**

TAKING IT FURTHER

- Why is metamorphic rock often used for sculptures and monuments? **It is usually very hard and durable. Nonfoliated rock has no crystal structure, so smoother edges are possible.**
- Why is metamorphic rock hard and durable? **The heat and pressure rearrange the crystal structure to form very strong bonds.**

CHALLENGE: METAMORPHIC MATCH WORKSHEET

_C_ Granite   _A_ Shale   _D_ Sandstone   _G_ Limestone
_B_ Mica   _F_ Basalt   _E_ Dolerite

LESSON 15

MINERALS

ANIMAL, VEGETABLE, OR MINERAL?

SUPPLY LIST

Copy of "Mineral Scavenger Hunt" worksheet

Supplies for Challenge: Egg carton   48 small rocks or pebbles

BEGINNERS

- How do you know if something is a mineral or not? **A mineral is solid, natural, not alive, has crystals.**
- Name at least three minerals. **Salt, sugar, silver, gold, copper, zinc, calcium, iron.**

MINERAL SCAVENGER HUNT WORKSHEET

1. Water: **Is not a solid.**
2. Steel: **Is man-made.**
3. Coal: **Is organic (made from plants).**
4. Cookies: **Are mixtures made from varying amounts of minerals.**
5. Glass: **Does not have a crystalline structure.**
- **Minerals around my house include: Calcium in your milk, iron in nails, talc in powder, zinc in your breakfast cereal and fluoride in your toothpaste, etc.**

WHAT DID WE LEARN?

- What five requirements must a substance meet in order to be classified as a mineral? **Naturally occurring, inorganic, constant chemical proportions, regular structure and solid.**
- What is a native mineral? **A mineral that has only one type of atom—a pure element.**
- What is a compound? **A substance with two or more elements in fixed proportions.**

TAKING IT FURTHER

- Are there any minerals that are mixtures? **No, mixtures do not have fixed proportions.**
- What is the difference between a rock and a mineral? **Most minerals are compounds and most rocks are mixtures. Also, rocks can have organic compounds and minerals do not.**
- Is coal a mineral? **No, it is an organic compound. It is a rock.**
- Are all minerals considered rocks? **In their naturally occurring state, minerals are considered rocks.**
- Are all rocks considered minerals? **No, most rocks are made of more than one kind of mineral.**
- Where are you likely to find minerals? **Just about everywhere, such as in your body, your food, toothpaste, coins, cars, radios, TVs, etc.**

IDENTIFYING MINERALS

IS IT SALT OR SUGAR?

SUPPLY LIST

Copy of "Mineral Identification" worksheet   Magnifying glass   Eye protection/goggles
Samples of 3 or 4 minerals (quartz, feldspar, mica, limestone, etc.)   Masking tape   Penny
Hammer   Old drinking glass   Old pillowcase or towel   Unglazed ceramic tile
Rocks and Minerals Guide
Supplies for Challenge: Copy of "Is it a Rock or a Mineral?" worksheet   Rocks and Minerals Guide

BEGINNERS

- What are three things you can look for to help you identify a rock or mineral? **Color, luster, and crystals.**

WHAT DID WE LEARN?

- What are some common tests used to identify minerals? **Color, streak, luster, crystal shape, hardness, cleavage.**

- Why is color alone not a sufficient test? **Many minerals have the same color. Also, the outside of a sample may change color when exposed to air or water.**

TAKING IT FURTHER

- Is crystal size a good test for identifying a mineral? Why or why not? **No, the size of the crystals is dependent on the temperature at which the sample formed, but is not as dependent on the type of material. Crystal shape is a much better test.**

- What is the difference between cleavage and fracture? **Cleavage indicates that a sample breaks smoothly in one or more directions—showing that the crystal structure is lined up in that direction. Fracture indicates that a sample breaks in smooth curves but not in straight lines.**

- Why do some tests need to be done in a laboratory? **Some tests are too dangerous to do at home and others require special equipment.**

- How can you tell a sample of sugar from a sample of salt? **You could taste the samples to determine which is sugar and which is salt if you are certain they are table salt and sugar. However, you should never taste an unknown sample. A better way would be to use a magnifying glass and observe the shape of the crystals. Salt crystals are usually perfect cubes and sugar crystals are longer and rectangular.**

CHALLENGE: IS IT A ROCK OR A MINERAL? WORKSHEET

Item	Element, compound, or mixture?	What are its main components?	Rock or mineral?
Gold	**Element**	**Gold**	**Mineral**
Granite	**Mixture**	**Quartz, feldspar, and mica,**	**Rock**
Feldspar	**Compound**	**Aluminum and silicon**	**Mineral**
Quartz	**Compound**	**Silicon and oxygen**	**Mineral**
Limestone	**Mixture**	**Calcium carbonate and other materials**	**Rock**
Copper	**Element**	**Copper**	**Mineral**

Diamond	Element	Carbon	Mineral
Obsidian	Mixture	Glass containing silica and aluminum	Rock
Gypsum	Compound	Calcium, sulfur and oxygen	Mineral
Shale	Mixture	Clay and mud	Rock

LESSON 17

VALUABLE MINERALS

HOW MUCH FOR AN OUNCE OF GOLD?

SUPPLY LIST

Chocolate chip cookies   Toothpicks

BEGINNERS

- Name four valuable minerals. **Gold, silver, copper, diamonds.**
- What is the hardest mineral? **Diamond.**
- What are some uses for valuable minerals? **Coins, jewelry, wires, film, pipes.**

WHAT DID WE LEARN?

- What are some valuable minerals? **Gold, silver, copper, and diamonds.**
- What is a native mineral? **A mineral made from just one kind of element.**
- What are some important uses for gold? **Jewelry, electronics, money.**
- What are some important uses for silver? **Jewelry, flatware, electronics, film processing.**

TAKING IT FURTHER

- Why is diamond considered an exception among minerals? **Minerals are not supposed to contain carbon—they are inorganic. However, diamond is pure crystallized carbon.**
- Diamonds and coal are both made from carbon. What makes them different? **Coal is a relatively soft black rock made from compressed plant material. Diamonds are carbon atoms that have crystallized due to extreme heat and pressure. Most geologists classify coal as a sedimentary rock and diamond as a metamorphic rock, although some geologists classify coal as a metamorphic rock as well.**

LESSON 18

NATURAL & ARTIFICIAL GEMS

CUT STONES

SUPPLY LIST

Copy of "Breastplate" worksheet   Colored pencils, markers, or crayons

Supplies for Challenge: Pictures of various gems   Visit a mineral exhibit at a museum (optional)

BEGINNERS

- What is a gem? **A stone that can be cut or polished to reflect light.**

- What are three common gems? **Diamond, ruby, emerald, sapphire, topaz, amethyst.**

- How are man-made gems different from natural gems? **They are not as strong; they are sometimes made from different materials; they are made by people instead of occurring naturally.**

WHAT DID WE LEARN?

- What is a gem? **A mineral that is valued by people because of its beauty. Usually gems are brightly colored and have perfect cleavage that allows light to reflect through them.**

- How is a gem different from a native mineral? **Native minerals have only one element. Diamonds are gems that are also native minerals. However, most gems are comprised of more than one element.**

- How are artificial rubies made? **By melting the elements that make natural rubies, then allowing the materials to slowly cool and crystallize.**

TAKING IT FURTHER

- What can you guess about the temperatures at which synthetic rubies are formed? **They are formed at high temperatures.**

- Why would rubies be formed at high temperatures? **Recall from lesson 10 that crystals grow larger when cooled more slowly, so rubies would be cooled at high temperatures to form larger crystals.**

- What are some disadvantages of synthetic gems? **They are not as durable; they are not as valuable; they sometimes look different than naturally occurring gems.**

- Why are natural gems worth more money than artificial gems? **Artificial gems are not as durable. Also, price is partially determined by availability. Natural gems are more scarce than artificial gems and are thus more expensive.**

QUIZ 2

ROCKS & MINERALS

LESSONS 8–18

Choose the best answer for each question.

1. _B_ What percentage of all fossils are fossilized dinosaur bones?
2. _A_ What rock is commonly used for buildings and monuments?
3. _C_ Which of the following is required in order for a plant or animal to fossilize?
4. _A_ What rock is made from the same element as diamonds?
5. _B_ Why should we be careful when using the results of Carbon-14 dating?
6. _C_ What is one common characteristic of metamorphic rock?
7. _A_ How many elements are in a native mineral?
8. _B_ What is a common mineral found in the human body?
9. _D_ Which of the following is a native mineral?
10. _D_ How are artificial gems easily identified?

Short answer:

11. Where is the earth's crust the thickest? **Earth's crust is thickest under the mountains.**

12. Name the three types of rocks. Describe how each type is formed and give an example of each.

    A. **Igneous—formed when melted rock cools. Examples: pumice, basalt, obsidian, granite.**

    B. **Sedimentary—formed from small bits of broken rock, shells and other materials that are cemented together, or when minerals precipitate out of water. Examples: sandstone, mudstone, shale, limestone, dolomite.**

    C. **Metamorphic—formed when igneous or sedimentary rocks are changed by pressure and heat over time. Examples: marble, slate, gneiss, quartzite.**

13. List three tests commonly used to identify minerals. **Tests to identify minerals include color, streak, cleavage, hardness, crystal shape, and luster.**

CHALLENGE QUESTIONS

Match the term with its definition.

14. _C_ The most common element in the earth's crust

15. _E_ Most common rock in continental crust

16. _G_ Most common rock in oceanic crust

17. _H_ Holes found in igneous rocks

18. _A_ Rock containing two or more sizes of crystals

19. _I_ Fragmental rock with rounded clasts

20. _B_ Fragmental rock with angular clasts

21. _J_ Fossilized animal dung

22. _D_ Mineral with atoms of only one type

23. _F_ Mineral with two or more elements in definite proportions

24. _L_ Order of rock layers according to evolutionists

25. _K_ Smooth stones found inside fossilized animal bodies

UNIT 3
MOUNTAINS & MOVEMENT

LESSON 19

PLATE TECTONICS

SLIP SLIDING AWAY

SUPPLY LIST

Copy of world map (lesson 5)   Scissors   Tracing paper   Tape
Supplies for Challenge: Graham crackers   Creamy peanut butter or frosting
Waxed paper

BEGINNERS

- What are large areas of land called? **Continents.**
- How many continents do scientists think existed before the Flood? **One.**
- What was that continent called? **Rodinia.**

WHAT DID WE LEARN?

- What is plate tectonics? **The theory that the crust is made up of several large floating plates.**
- How many plates do scientists think there are? **13 plates: 6 major, 7 minor.**

TAKING IT FURTHER

- What are some things that are believed to have happened in the past because of the movement of the tectonic plates? **Continents separated by sea-floor spreading, mountains and ocean basins were formed, volcanic activity.**
- What are some things that happen today because of the movement of the tectonic plates? **Earthquakes, volcanic activity, forming of faults and rifts.**

LESSON 20

MOUNTAINS

DON'T MAKE A MOUNTAIN OUT OF A MOLE HILL

SUPPLY LIST

Copy of "Famous Mountains" worksheet

Supplies for Challenge: Copy of world map   Atlas or topographical map of world

BEGINNERS

- What is a mountain? **An area of land that is taller than the land around it.**

- What is the tallest mountain in the world? **Mount Everest.**

- Name one famous mountain range. **Cascades, Sierra Nevadas, Rocky Mountains, Appalachian Mountains.**

FAMOUS MOUNTAINS WORKSHEET

- _C_ Mountains of Ararat

- _E_ Mount Sinai

- _G_ Mount Moriah

- _D_ Mount Carmel

- _F_ Mount of Olives

- _A_ Mount of Transfiguration

- _B_ Mount Horeb (The Mountain of God)

WHAT DID WE LEARN?

- What is a mountain? **A rise in land with steep sides going up to a summit.**

- What is a mountain range? **A series of mountain peaks in a given area.**

- What is the difference between actual height and elevation of a mountain? **Actual height is the difference between the base and summit of a mountain. Elevation is the height of the summit above sea level.**

TAKING IT FURTHER

- Where are the mountains with the highest elevations located? **In the Himalayan Mountain system in the area between India and China.**

- Is a 700 foot rise a mountain or a hill? **It depends on your perspective. In areas with 5,000 foot mountains, a 700 foot rise will probably be called a hill, but in a relatively flat area it may be called a mountain.**

LESSON 21 — TYPES OF MOUNTAINS

HOW DID THEY FORM?

SUPPLY LIST

Newspaper or paper towels
Supplies for Challenge: Sponge

BEGINNERS

- Describe three ways that new mountains can be made. **When material is deposited by a volcano or water, when surrounding material is worn away by water or wind, when the earth's crust is pushed on from two sides.**
- What is the likely cause of many of the mountains we see today? **The Genesis Flood.**

WHAT DID WE LEARN?

- How are depositional mountains formed? **Debris such as ash, lava, sand, etc. is deposited over time, eventually forming a mountain.**
- How are erosional mountains formed? **Large amounts of material are eroded away, such as by a flood, leaving behind mountains and valleys.**
- How are fold and fault mountains formed? **Tectonic plates push against each other and the constant force causes rocks in the middle to fold or slip and push up to form mountains.**

TAKING IT FURTHER

Identify each mountain as either depositional, erosional or fold:
- Mount St. Helens: **This is a volcano—depositional.**
- Bryce Canyon: **Area with many flat-topped sandstone mountains—erosional.**
- Sand Dunes National Monument: **Sand deposited by water and wind—depositional.**
- Rocky Mountains: **Large mountain range—fold.**
- Grand Canyon: **Believed to be carved by a flood—erosional.**
- Mount Everest: **Highest mountain on earth—fold.**

LESSON 22 — EARTHQUAKES

SHAKE, RATTLE, AND ROLL

SUPPLY LIST

10–20 building blocks
Supplies for Challenge: Three different colors of modeling clay

BEGINNERS

- How often do earthquakes happen? **Several times each day.**
- What is an aftershock? **A smaller earthquake that occurs after a large earthquake.**
- What is a tsunami? **A giant wave that can be triggered by an earthquake.**

EARTHQUAKE-PROOF BUILDINGS

- What might architects do to help make buildings stronger? **Use reinforcing materials like rebar, overlap joints, and use strong materials.**
- What shape of building is more likely to withstand an earthquake? **Short, broad buildings generally do better than tall, narrow designs.**

WHAT DID WE LEARN?

1/16 review

- What is believed to be the cause of earthquakes? **Tectonic plates move against each other causing stress or strain on the rocks. When the stress becomes too great, the rocks move quickly, resulting in an earthquake.**
- What is an aftershock? **A smaller quake that occurs after a major earthquake.**
- What name is given to the area on the earth's surface above where an earthquake originates? **Epicenter.**
- What is a fault? **A crack in the earth's crust resulting from an earthquake.**

TAKING IT FURTHER

- How does the type of material affect the speed of the earthquake waves? **Earthquake waves move more quickly through rock and other dense materials. The waves slow down when traveling through sand, mud and liquid rock or magma.**
- How does this change in speed help scientists "see" under the earth's crust? **Scientists can track the speed of earthquake waves under the crust. When the waves change speed, this tells the scientists that the material at that location is a different density. By tracking this, scientists can predict the thickness of the crust and the density of the magma below it.**
- Why are earthquakes in the middle of the ocean so dangerous? **They trigger tsunamis that can kill people hundreds of miles away.**

LESSON 23

DETECTING & PREDICTING EARTHQUAKES

PREDICTING THE "BIG ONE"

SUPPLY LIST

Shoebox   Rolling pin   Paper   Pencil   Tape
Supplies for Challenge: Copy of world map

BEGINNERS

- What instrument is used to detect earthquakes? **A seismograph.**
- How can people make buildings safer during earthquakes? **Make them stronger so they can withstand the shaking.**

- Can people accurately predict when and where an earthquake will happen? **No, only God knows when and where earthquakes will happen.**

WHAT DID WE LEARN?

- What is the difference between the magnitude and the intensity of an earthquake? **Magnitude measures the actual strength of the earthquake—how strongly it moved the earth; intensity describes the damage done by the earthquake.**

- What are three factors that determine how much damage is done by an earthquake? **Where it occurs, how strong it is, how long it lasts, how the buildings are built.**

- Explain how a seismograph works. **The part with the rotating drum moves with the earth, and the part with the pen or light is attached to a mass that does not move with the earth.**

- What people group was first to record earthquake measurements? **The Chinese about AD 132.**

TAKING IT FURTHER

- What are some ways people have learned to prepare for earthquakes? **Buildings in areas that are prone to earthquakes are designed and built to be able to withstand vibrations. Also, scientists have an extensive network of seismographs and other instruments to detect earthquakes in hopes of giving enough advance warning for people to leave the area, although this has not proven very effective.**

- What should you do if you are in an earthquake? **The best thing to do is get under something strong, like a sturdy table that can protect you from falling debris.**

LESSON 24
VOLCANOES
FIRE MOUNTAINS

SUPPLY LIST

Baking soda   Vinegar   Empty bottle   Newspaper   Tape   Baking sheet
Red food coloring (optional)
Supplies for Challenge: World map from previous lessons   World atlas

BEGINNERS

- What causes a volcano to erupt? **Pressure under the earth from plates rubbing against each other.**
- What is lava? **Melted rock.**
- What can come out of a volcano besides lava? **Ash, cinders, gases, and bombs.**
- What is the "Ring of Fire"? **The area around the Pacific Ocean where most of the volcanoes exist.**

WHAT DID WE LEARN?

- What are the three stages or states of a volcano? **Active, dormant, extinct.**
- Describe the three main parts of a volcano. **Magma chamber—area below crust filled with melted rock; central vent—channel through which magma forces its way to the surface; crater—indented area around the mouth of the volcano where a cone formed by solidified ash and lava has collapsed.**
- Give the name for each of the following items that are emitted from a volcano:

1. Liquid or melted rock: **Lava.**
2. Tiny bits of solid rock: **Ash.**
3. Pieces of rock from 0.2 to 1 inch (0.5–2.5 cm) in diameter: **Cinders.**
4. Blobs of lava that solidify in the air: **Bombs.**
5. Steam and carbon dioxide: **Gases.**

TAKING IT FURTHER

- How might a volcano become active without anyone noticing? **If it is underwater or located in a very remote area.**
- How are volcanoes and earthquakes related? **Both primarily occur where two tectonic plates meet and cause pressure to build up.**
- How certain can we be that a volcano is really extinct? **Not completely. Several volcanoes have become active after being classified as extinct.**

LESSON 25

VOLCANO TYPES

read pg 111 112

IS THERE MORE THAN ONE?

SUPPLY LIST

Pie pan   Cookie crumbs   ½ gallon of ice cream   Chocolate chips   Chocolate syrup
Supplies for Challenge: Unopened can of soft drink

BEGINNERS

- What kind of mountain is formed when mostly lava comes from a volcano? **Shield.**
- What kind of mountain is formed when mostly ash and cinders come from a volcano? **Cinder cone.**
- What kind of mountain is formed when a volcano alternates between lava and solid material? **Composite.**

WHAT DID WE LEARN?

- What are the three shapes of volcanoes and how is each formed? **Shield—mostly from lava; Cinder cone—mostly from solid debris such as ash and cinders; Composite—alternating between lava and solid matter.**
- Where are most active volcanoes located today? **Along the Ring of Fire around the Pacific Ocean.**
- What are some of the dangers of volcanoes? **Fire, debris, suffocation from gases, mudslides, tsunamis.**
- What are some positive side effects of volcanoes? **Geothermal energy, sulfur deposits, fertile soil, new land, heat vents in the ocean, black sand beaches.**

TAKING IT FURTHER

- How is the formation of black sand beaches different from the formation of white sand beaches? **Black sand is formed when hot lava shatters as it cools very suddenly when reaching the ocean. Most white sand beaches are composed of crushed shells.**

LESSON 26

MOUNT ST. HELENS

GOD'S GIFT TO SCIENTISTS

SUPPLY LIST

Copy of "Volcano Word Search"

BEGINNERS

- What happened at Mount St. Helens on May 18, 1980? **It erupted with a huge explosion.**
- How long did it take for 25 feet of layered ash to accumulate after the eruption? **Only one day.**
- What did the flood of mud and water from the volcano do? **It carved a canyon that was 100 feet deep.**

VOLCANO WORD SEARCH

```
A D O P X Y B A E P A H D S H
C O D O R M A N T E R A C T C
E X S N M K L A I O T C A E R
P A H O E H O E A S M A G R O
U U R S I N D R C H A P T U O
C U S U B D U C T I O N P P Q
O A U L I N S I I E E C R T E
M A G M A O M N V L X Y Z I O
P L M K R V Z D E D T U X O P
O S W Q B O A E L I I F F N A
S C L R E T A R C A N P O S T
I T S U I O B O M B C S H I L
T C N E D R N H F R T W K L J
E C A L D E R A E T X N T M I
R W I O P R Q C M I O R S L P
```

WHAT DID WE LEARN?

- Describe some of the ways the data collected at Mount St. Helens is challenging evolutionary thinking. **Canyons formed in only one day, 25 feet (7.6 m) of sedimentary layers laid down in only one day, upright trees on the bottom of Spirit Lake explain "petrified forests."**

TAKING IT FURTHER

- How did the ash from the eruption of Mount St. Helens affect the weather in 1980? **It darkened the skies and cooled the temperatures, not just near the eruption site, but around the world.**
- How could volcanic activity have contributed to the onset of the Ice Age? **Massive amounts of ash in the atmosphere would have blocked out much of the sunlight, causing cooler summers. Cooler temperatures combined with the large amount of water vapor from warm oceans would cause accumulation of snowfall leading to glacier formation.**

MOUNTAINS & MOVEMENT

LESSONS 19–26

Match the term with its definition.

1. _F_ Theory that the crust is composed of several large landmasses

2. _C_ Name given to original landmass

3. _H_ Series of mountain peaks in a given area

4. _I_ Highest mountain peak on earth

5. _E_ Center of earthquake activity

6. _L_ Smaller quakes after a major earthquake

7. _N_ The Richter scale measures this

8. _D_ Instrument for measuring earthquakes

9. _O_ Blobs of lava that harden in the air

10. _B_ A volcano that has not erupted in the past 50 years

11. _M_ A volcano that is not expected to erupt again

12. _G_ Volcano that recently erupted in Washington state

13. _J_ Volcano that had one of the largest eruptions ever

14. _K_ Type of volcano formed from lava and solid material

15. _A_ One tectonic plate sliding under another

CHALLENGE QUESTIONS

Mark each statement as either True or False.

16. _T_ Continental drift is the name given to the movement of tectonic plates.

17. _F_ Rifting occurs when two tectonic plates collide.

18. _F_ There are relatively few mountain ranges in the world.

19. _T_ Pressure on tectonic plates can cause rocks to fold or bend.

20. _T_ An anticline is formed when rocks bend upward.

21. _F_ A hanging wall is the rock layers below a fault.

22. _T_ A hanging wall moves downward in a normal fault.

23. _F_ Most earthquakes and volcanoes are located around the Atlantic Ocean.

24. _F_ Basaltic volcanoes usually have steep sides.

25. _T_ Scientists cannot accurately predict when a volcano will erupt.

Unit 4
WATER & EROSION

LESSON 27
GEYSERS
HEATED GROUND WATER

SUPPLY LIST

Flexible soda straw   Cup filled with water

BEGINNERS

- What causes a geyser to erupt? **Pressure from water that is heated underground.**
- How is a hot spring different from a geyser? **The water in the pool just bubbles up, but the water in a geyser shoots out.**
- Where can you see many geysers? **Yellowstone National Park.**

WHAT DID WE LEARN?

- What are some ways that heated ground water shows up on the surface of the earth? **Hot springs and pools, spouters, fumaroles, mud pots, mud volcanoes and geysers.**
- Explain how a geyser works. **Underground water is heated and expands inside a network of "plumbing." When the pressure of the heated water becomes greater than the weight of the water above it, it forces water up through the vent.**
- How is a mud pot different from a hot spring? **A mud pot has more dirt than water in it, whereas a spring is mostly just water.**

TAKING IT FURTHER

- How might a scientist figure out which irregular geysers are connected underground? **By observing the behavior of the geysers. If one becomes active at the same time another becomes inactive they might be connected.**
- Why do some hot pools have a rainbow appearance? **The temperature of the water cools as it spreads out and different colored algae and bacteria grow in different temperatures of water.**
- Can you tell the temperature of the water just by looking at a pool? **Maybe. Green colored bacteria begin to grow in water just below 167° F (75° C). Other colors grow in lower temperatures; however, this may not be a completely accurate way to determine temperature.**

CHALLENGE: GEOTHERMAL ENERGY

- Why are geothermal power plants mostly located near edges of tectonic plates? **Because that is where magma is most likely to find its way close to the surface of the earth.**

- Would you expect geothermal power plants to experience more or fewer earthquakes than other power plants? **In general, you would expect more, because they are built in areas that have moving tectonic plates and are more likely to have earthquakes.**
- Why is geothermal energy considered a renewable resource? **The magma and water are not being used up so there is a constant supply of steam. Even though water is "lost" to evaporation, it eventually condenses and returns to the ground through precipitation.**

LESSON 28

WEATHERING & EROSION

IT'S WEARING ME DOWN

SUPPLY LIST

Copy of "Weathering" worksheet   Vinegar   Bar of soap   Soda straw
Real chalk (made from limestone) or a limestone rock   Modeling clay
Supplies for Challenge: Copy of "Chemical Erosion" worksheet   Steel wool (without soap)
3 plastic zipper bags

BEGINNERS

- What is erosion? **When rocks are worn away a little at a time, also called weathering.**
- What happens to a rock when water freezes inside a crack? **The crack gets bigger and eventually the rock will break.**
- What are three ways that rocks can be eroded? **Freezing water, wind, plant roots, chemicals.**
- Does erosion happen slowly or quickly? **Usually it happens slowly, but it can happen quickly during a flood.**

WEATHERING WORKSHEET

- What did you observe as the vinegar came in contact with the limestone? **It should have fizzled.**
- How did the limestone look afterward? **Some of the limestone should have been worn away.**
- How did the dripping of the water affect the surface of the soap? **It should have cause a pit in the soap.**
- How did the clay and straw change when the water froze? **The water expands as it freezes, either pushing out the clay, or breaking the straw.**
- What effects might freezing water have on rocks? **As water freezes it expands, widening cracks and breaking rocks apart.**

WHAT DID WE LEARN?

- What is weathering? **The wearing down of rocks by natural forces.**
- Describe the two types of weathering. **Chemical—material is changed by chemical reactions; Mechanical—material is worn away by pressure from water/ice, wind, debris, plant roots.**

TAKING IT FURTHER

- How does freezing and thawing of water break rocks? **Water expands when it freezes, breaking off bits of rock and enlarging the crack. After many cycles, the rock will break.**

- What do people do that is similar to mechanical weathering? **People use sand blasting to remove graffiti. Strip mining and dynamite are used to remove rock.**

CHALLENGE: CHEMICAL EROSION WORKSHEET

- Why do you think this sample had the most rust? **Water removes (erodes) the rust from the iron and allows the air in the bag to react with more iron causing more oxidized iron atoms. The bag with water and no air had some rust because of small amounts of oxygen in the air left in the bag and dissolved in the water.**

LESSON 29

MASS WASTING

THE FORCE OF GRAVITY

SUPPLY LIST

Baking tray or large baking pan   Soil and rocks

BEGINNERS

- What force pulls dirt and rocks down a hill? **Gravity.**
- What can trigger a landslide? **Heavy rains or an earthquake.**
- What is an avalanche? **When ice and snow moves quickly down a hill.**

WHAT DID WE LEARN?

- What is mass wasting? **The movement of large amounts of rocks and soil due to gravity.**
- What is slow movement of the soil and rocks down a slope called? **Creep.**
- What is rapid or sudden movement of the soil and rocks called? **A landslide.**

TAKING IT FURTHER

- How does water affect mass wasting? **Water can loosen the bonds between the rocks and soil, allowing gravity to move them more easily.**
- How might weathermen predict when the avalanche danger is high? **The danger might increase when snow suddenly builds up during a storm, when winds are high or when the temperatures begin to warm up. All of these factors can change the strength of the bonds of the snow on the side of the mountain. However, weathermen cannot accurately predict when and where a specific avalanche will occur.**

LESSON 30

STREAM EROSION

THE POWER OF MOVING WATER

SUPPLY LIST

3 baking pans   Soil   Leaves, grass, or other plant material   Book at least 2 inches thick
Supplies for Challenge: Soil   3 paper cups   Pencil   Acces to an oven

Beginners

- What is stream erosion? **Wearing away of rocks and soil by moving water.**
- Why does water flow downhill? **Gravity pulls it down.**
- Does water flow slowly or quickly down a steep hill? **Quickly.**

What did we learn?

- What is the most powerful eroding force? **Moving water.**
- How does gravity cause stream erosion? **Gravity pulls water downhill. The steeper the hill, the faster and more powerfully the water flows.**
- What is the gradient of a river? **The difference in elevation between the headwaters, or source, and the mouth, or lowest point.**

Taking it further

- Why are farmers concerned about soil erosion? **Topsoil is difficult to replace, so it must be protected to enable the farmers to grow good crops.**
- What are some steps farmers take to prevent water from eroding their topsoil? **They plow crossways to the flow of the water; they alternate crops; they terrace steep areas.**
- Besides water, what other natural force can erode topsoil? **Wind can blow it away.**
- What can farmers do to protect their topsoil from wind erosion? **They can plant trees to block or break up the wind. They can plant shorter rows to keep the wind from blowing soil too far away.**
- Why do lakes and reservoirs have to be dredged, emptied and dug out periodically? **Water from incoming streams deposits silt and rocks in the bottom of the lake and eventually fills it up.**

LESSON 31

SOIL

Isn't it just dirt?

Supply list

Potting soil   Soil from your yard   Magnifying glass
Supplies for Challenge: Copy of "Permeability of Soil" worksheet   4 paper cups
2–3 cups of soil   Newspaper   Colander   Fine mesh strainer   Stop watch
Liquid measuring cup   Pencils   Baking sheet

Beginners

- What is soil made of? **Bits of sand, clay, silt, and dead plants and animals.**
- Where do bits of sand and clay come from? **They are broken off of rocks by erosion.**
- Is erosion always bad? **No, it helps to make new soil.**

What did we learn?

- What are the major components of soil? **Sand, silt, clay, organic material (humus).**
- What is the most important element in soil for encouraging plant growth? **Humus—decayed plant matter.**

TAKING IT FURTHER

- What type of rocks would you expect to find near an area with sandy soil? **Quartz rocks.**

- What type of rocks would you expect to find near an area with clay soil? **Feldspar and mica.**

- How does a river that regularly floods, such as the Nile, restore lost topsoil? **Soil is washed into the river by the moving water. When it floods, much of this soil is moved out of the riverbed to the area around the river. As the floodwaters recede, the soil is left behind and can be used for farming.**

- What are some ways that farmers restore nutrients to the soil? **Chemical fertilizers, animal waste, rotating crops, plowing under crops.**

CHALLENGE: PERMEABILITY OF SOIL worksheet

- Which sample had the highest permeability? **Sample 2 should have the highest permeability because it has the largest particles.**

- Which had the lowest permeability? **Sample 4 should have the lowest permeability because it has the smallest particles.**

- How did your unsifted sample (sample 1) compare to the sifted samples? **Sample 1 should be somewhere in between since it is a combination of the other three.**

- How does particle size affect permeability? **The larger the particles the greater the porosity and thus the greater the permeability. In other words, the bigger the particles the more air space there is so the faster the water can flow through.**

LESSON 32

GRAND CANYON

LOTS OF TIME, LITTLE WATER OR LOTS OF WATER, LITTLE TIME?

SUPPLY LIST

Modeling clay   Paper   Markers

BEGINNERS

- What are the two main ideas for how Grand Canyon was formed? **A little water over a long time or a lot of water over a little time.**

- Which idea agrees with the Bible? **A lot of water over a little time.**

- Where could a lot of water have come from to make the canyon? **Noah's Flood.**

WHAT DID WE LEARN?

- What is the main controversy between evolutionists and creationists concerning the formation of Grand Canyon? **Was it formed by a little water over a long period of time, or by a lot of water over a short period of time?**

- What evidence shows radiometric dating methods to be unreliable? **Rocks at lower levels were dated younger by 100s of millions of years than rocks at higher levels.**

TAKING IT FURTHER

- What event at the eruption of Mount St. Helens supports the biblical view of how Grand Canyon was formed? **A flood resulting from the eruption formed a canyon 100 feet (30 m) wide and 100 feet (30 m) deep through solid rock in only one day. A flood on a much larger scale would have much larger effects.**

- How can scientists look at the exact same data and draw different conclusions? **Everyone has preconceived ideas, or presuppositions, that affect how he or she views the evidence. Evolutionists believe the earth is billions of years old, so they reject any theories that conflict with that idea. Creationists believe what the Bible says and interpret the evidence from that point of view.**

- How can we know what to believe when scientists disagree? **We must trust what God says.**

LESSON 33

CAVES

UNDERGROUND WONDERLANDS

SUPPLY LIST

2 paper or plastic cups   Cotton string   Epsom salt   Cardboard
Supplies for Challenge: Research materials on caves

BEGINNERS

- What is the name given to the rocks that form when chemicals build up on the ceiling of a cave? **Stalactite.**

- What is the name given to rocks that form when chemicals build up on the floor of a cave? **Stalagmite.**

- How are these special rocks formed? **When water passes through limestone it dissolves chemicals that are then left behind in the caves after the water evaporates.**

WHAT DID WE LEARN?

- How are the beautiful formations in caves formed? **Calcite is dissolved in water as it passes through the limestone. As the water evaporates, it leaves the calcite behind, forming beautiful shapes inside the caves.**

- What is a stalactite? **A formation in a cave that hangs from the ceiling.**

- What is a stalagmite? **A formation in a cave that forms on the floor.**

TAKING IT FURTHER

- What evidence do we have that formations in caves can develop rapidly? **Formation has been measured to be rapid in some wet caves. Also, bats and other animals have been found preserved in some stalactites and stalagmites.**

- Why is it likely that calcite formations would have formed rapidly after the Flood? **Conditions would have been very wet inside the caves, with many minerals suspended or dissolved in the water.**

- Besides in caves, where can calcite deposits be found? **Near geysers in places such as Yellowstone National Park.**

QUIZ 4

WATER & EROSION

LESSONS 27–33

Fill in the blank with the correct term from below.

1. A thermal feature that shoots hot water many feet into the air is a _geyser_.

2. _Old Faithful_ is one of the most famous geysers in the world.

3. _Geothermal_ energy can be obtained from areas containing geysers.

4. Geysers often contain _hydrogen sulfide_, which gives them a bad smell.

5. A _fumarole_ is produced when super-heated steam reaches the surface.

6. The process of wearing down rocks is called _weathering_.

7. _Frost heaving_ is the process that brings rocks to the surface each winter.

8. _Mass wasting_ is the effect of gravity pulling soil and rocks down a hill.

9. Rapid movement of large amounts of rocks and soil is called a _landslide_.

10. The most powerful eroding force is _moving water_.

11. The most important component of soil for growing plants is _humus_.

12. _Ash_ can become fertile soil after a volcanic eruption.

13. A formation in a cave that goes from floor to ceiling is called a _column_.

14. _Gravity_ is the force that causes water to move rapidly down a hill.

15. _Limestone_ is the main type of rock from which caves are formed.

CHALLENGE QUESTIONS

Short answer:

16. Where is the most likely place to find geothermal areas? **Near the boundaries of tectonic plates.**

17. List two types of chemical erosion. **Acids, water, oxidation.**

18. Which is more easily eroded, iron or rust? **Rust.**

19. Where are rock glaciers likely to be located? **Steep mountain slopes with cool summers (Colorado and Alaska).**

20. What is a fossil rock glacier? **A rock glacier that no longer has any ice in it.**

21. Which has more power to erode, fast moving water or slow moving water? **Fast moving water.**

22. What is porosity in soil? **The measure of the amount of air space in the soil.**

23. What is permeability of soil? **The speed at which water flows through a sample of soil.**

24. Why are porosity and permeability important? **They affect how well plants will grow in the soil.**

25. What are two indications of a large scale flood found in Grand Canyon? **Two hundred miles of level sedimentary layers, abundant aquatic fossils, nautiloid fossils all lined up the same way, fossilized amphibian and reptile footprints in sandstone.**

26. What is the most interesting thing you learned about caves? **Accept reasonable answers.**

LESSON 34

ROCKS & MINERALS COLLECTION

PUTTING IT ALL TOGETHER

FINAL PROJECT SUPPLY LIST

Rocks and Minerals Guide   Samples of rocks, gems, and minerals   Glue   Markers
Tagboard or poster board   Display box with separated sections (optional)

What did we learn?

- What are the three types of rocks? **Igneous, sedimentary, and metamorphic.**
- What is a native mineral? **One made from a pure element such as gold or silver.**

Taking it further

- What are some of the greatest or most interesting things you learned from your study of our planet earth? **Accept reasonable answers.**
- Read Genesis chapters 1 and 2. Discuss what was created on each day and how each part completes the whole. **Consider using the drawing lesson at www.answersingenesis.org/docs2002/oh20020301_112. asp to develop this topic.**
- What earth science topic would you like to learn more about? **Have students research at the library or the AiG website and the Internet.**

Our Planet Earth

Lessons 1–34

Label this diagram of the earth.

A. **Crust**   B. **Mantle**   C. **Outer Core**   D. **Inner Core**

Mark each statement as either True or False.

1.   _T_ Scientists "see" beneath the crust of the earth using earthquake (seismic) waves.
2.   _F_ The core is the coolest part of the earth.
3.   _T_ Magma is liquid rock under the surface of the earth.
4.   _F_ Coal is not a rock because it is organic.
5.   _F_ Rocks never change form.
6.   _T_ Rocks are made from one or more minerals or organic materials.
7.   _T_ Larger crystals form in igneous rock if it is cooled slowly.
8.   _F_ All rocks sink in water.
9.   _T_ Sandstone is a sedimentary rock.
10. _T_ Limestone frequently contains fossils.
11. _F_ Fossils prove that evolution is true.
12. _F_ Fossils prove that creation is true.
13. _T_ Natural gas is almost always found near oil deposits.
14. _F_ Metamorphic rock can form at low temperatures.
15. _F_ Color is the most accurate way to identify a mineral.

Fill in the blank with the correct term from below.

16. The _**1st law of thermodynamics**_ states that matter cannot be created or destroyed.
17. The _**2nd law of thermodynamics**_ states that all systems tend toward a state of chaos.
18. Most _**evolutionists**_ believe that everything in nature happened only by natural processes.
19. _**Uniformitarianism**_ is the belief that everything was formed by the slow processes we observe today.